"Save Us Or Damn Us, Your Choice"

A Climate Custodian Publication

The Climate Covenant

TheHuMan is the avatar used by the author to explore this most serious subject with others all around the world over many, many years. His knowledge of this subject is wide and spreads across climate, politics and finance. His intent is not just to write books, but to create solutions that work for all of us.
This is the first book of this trilogy. Enjoy it.

A Climate Custodian Publication

Published By The Climate Custodian
For Everyone

"The author wishes to thank the many souls whose efforts
in laboratories, in the wilds and most notably those
anonymous souls across the internet who have given
so much time and energy to this cause and have truly been
on the front-line of this battle for whom
any acknowledgements or thanks are few and far between.
Thanks to you all."

To my family for supporting me,
To my friends for always being there,
To everyone else, for reading.

Internet Links
www.theclimatecovenant.com
www.climatetrilogy.com
www.climatecustodian.com

Media
@ClimateCovenant www.facebook.com/climate.covenant
@ClimateTrilogy www.facebook.com/climate.trilogy
@ClimateCustodia www.facebook.com/climate.custodian

Photos
All text, photos & illustrations produced by the Author

ISBN 978-0-9927887-0-4

First Edition
v787

Copyright @ Garry Coventry Higgs, December 2013

All rights reserved. No part of this book may be reproduced or transmitted in any form or by any means, electronic or mechanical, including photocopying, recording, or by any information storage and retrieval system, without the written permission of the author except where permitted by law.

The Climate Covenant

TheHuMan
We Need The Planet, The Planet Does Not Need Us

Climate Custodian

The Climate Covenant

Setting The Scene
1. A Curious Beginning to a Grave Problem 10
2. Introducing the Climate Covenant 20
3. The Climate Covenant 25

Psychology
4. Death 30
5. Compassion & Conscience 33
6. Conformity 36

The CO2 Trend
7. CO2, 2.2% & the Shadow... 40
8. Why Understand the CO2 Trend? 49
9. The Big Lag 51

CO2 Consequences
10. Breathing the Air 56
11. Ocean Acidification 62
12. Warming 67
13. Sea Rise 75
14. More Fossil Fuelled Consequences 80

Answering The Climate Covenant
15. Choosing Your Climate Covenant 88
16. Covenant Table 94
17. When Should We Start? 95
18. Making Up Your Mind With A Quick Recap 103
19. Register Your Climate Covenant 109

Thank You, Good Reading & Good Luck

Setting The Scene

Section 1

1 A CURIOUS BEGINNING TO A GRAVE PROBLEM

You probably picked this up when no-one was looking, a quick glance over your shoulder and a stealthy strafe of the shelf to accidentally grasp the unexpected victim.

You are perhaps feeling somewhat uneasy already and are just waiting for your moment to return it, hopefully unnoticed.

Trying to respect the ideal that one should not return the book too quickly else be seen as being too impulsive, rude or arrogant but has now not picked up the book in total error, you take just enough time to glance at the contents, browse the back cover or flick aimlessly through a few non-descript pages perhaps pausing at the odd stirring picture.

Quietly slipping the book back, you are relieved that no-one has noticed other than the half-asleep poodle and her loosely clad owner whom you are relieved hasn't moved to speak to you as you had first feared.

You are a little annoyed at yourself, how dare you stoop to the level of a bandana-wearing, green, tree-hugging, liberal hippy, even for a few minutes. Your public image could never stand that kind of stain. Very soon you will own a camper, be sitting in a damp field with tanned, long-haired warblers drinking herbal tea and pushing half-burnt tree limbs back into the fire.

Your name will have changed to Lovechild and you will be ostracised from all your normal social circles forthwith with no possibility of redress.

You have clearly lost your mind and shall remain so until your death when you will be fertilised along with your sandals to be reborn as a bushel of wheat. Later to be eaten by a spotty teenager thus completing the circle of life.

All this from your choice to read on only to end up having to spurn the superior choice of travelling half-dead on the train riddled with greed and ignorance, chubby in the behind and fuzzy in the head clutching the latest tabtop, dreaming of the annual two week break on the Costa Clubba where one can forget the other 50 weeks by sitting on a plastic chair and listening to urban melodies whilst sipping some green and blue meths. To settle down to read about another celebrity life story that was picked up randomly at the airport.

Thus to continue until dying from a heart attack a few years later from eating too much shovelled bacon in the office canteen between meetings over the facilitation of the corporate ethos and how long to stretch the meeting out in order to go straight home on the 5 o-clock train. Oh, how much joy we share.

Beauty and happiness are in the eye of the beholder. Our world is beautiful, at least for now. But we can be ugly. We can be selfish, cruel and foolish. Are we being that now? What good is money in a dying world? Will we smile when fishing from the top floor whilst ignoring our dead? Maybe I jest, maybe I do not. We are here to find out. The truth is the most powerful and painful of things, so it is wise to treat truth with respect.

Of course, we all have to do what we have to do to support our families. We quietly conform when society chooses the delights of the service sector for us. If social tradition dictates that tweed must be worn on a Sunday to the shoot then who are we to go against the tide in such an open display of disrespect and outright idiocy.

Obviously never a thought for the poor blighters chased out of their nesting areas only to be maimed on their return, chewed by a dog before having their necks wrung. Not really sporting. It is the fashion that counts though, don't you know. We think of ourselves as being one thing, when the reality is quite another.

Typically happily, as a little boy, I once enjoyed building camps, splashing about in streams, crawling about in the grass, thrashing the nettles with a big stick and trying to tease fish with a bit of gnarly old string. I remember hot summers, crisp springs and cold winters.

Autumn was a time for playing in the leaves and wondering what was happening to the daylight and who took away the evenings.

I remember wondering how babies were born and what were girls for? What made the snow and where were the birds always going? The world was full of wonder and beauty and I loved every moment that I spent within it. All this is being taken for granted. All of this is under threat by you and me, right now.

You may think that this is leading to saying that modern society is all at fault. It isn't, even if some choices are. All that is being asked is to think more and follow less. We are here, no-one else. It is up to us. This is our responsibility and that of no others.

It is the human mind that is the first thing that we will consider. It is often not a very logical place. And at worst, it can be a very immoral place, devoid of responsibility or care for what we do.

Consider things that have been or are being done in the name of law and power, even by those that never really agreed anyway.

I can kill in war as long as I do not see his face or meet him first. Kill her, she is a witch. He is a slave, look at his skin. Kill the child, she isn't one of us. Crucify him, his words offend us.

Control them, they are women. Forget them, they are fathers. Shoot them, friend. Is it really true that we cannot care more?

These are not all yesterday's mistakes. Do not be fooled into making the same mistake of blindly accepting law or society. The common man is capable of atrocities even if he knows it is wrong. As you will find out, a lot of bad things are coming and it is only you that can stand in the way, if you so choose.

Learning from past mistakes does not always happen. Life is confusing. Mistakes are made. Things are not always obvious.

Laws are only the point of view of one or more people at one time for one purpose.

Just being a law does not automatically make it right or moral. It is often only history that has the sight to correct mistakes; we are making them now. We do not have time to just wait for hope to prevail.

To avoid the biggest mistake in history we need to act, but when? Today, last year or last century? You will decide. We are doing nothing and very bad things are already on their way with far worse to come. Yes, that does include the risk of human extinction.

I wish a little tree-hugging, tea making and songs could solve all of this. They can't. We need the majority to stand up and be heard. And yes, that includes you. It includes all of us.

The truth is that our generation, regarding our environment, could not be more wrong. We are a tragic, selfish generation who will cost future generations a lot of lives, possibly all lives. Maybe Earth exists far away from other life for a reason. Perhaps whoever designed this place was not willing to have us destroy others the way we destroy ourselves. If it were me, I wouldn't.

We know death. This is on an entirely different scale against an even more defenceless foe; those yet to be born. Is it funny to harm and kill all humans yet to be born? We may not like it, but we value ourself, what is in it for myself and how much money we have above all else. We start to care for others or we all die.

We like to think of ourselves as modern when we are no different in constitution and thought to those that came before us. We seem to keep making the same mistakes for the same reasons; greed, vanity, money, power and so on. It is time to grow up.

These are disparaging remarks but they are not meant to offend. Don't be offended. Don't make that excuse. Facing this is hard. The Covenant will be hard if you answer it carefully. Deal with that answer on the facts alone. There's no time for complacency nor doubt. Nor for lazily believing disinformation. This book contains the facts. The level of morality though is left to you.

You will find out that there is undoubtedly a grave problem that we are responsible for. There is no excuse and that may upset you. What is important is what you do. Will you help or not?

At this point in the book, I can simply tell you that this is serious, that together we can solve it and that if you want to disagree then you have to have the foresight and strength to work out and express why. This is no joke. This is not playtime. Bad things are going to happen because of us and we need to think.

Perhaps we should remember what we are fighting for? To keep our only planet healthy so that we may survive upon it. There is nothing more important. Life now is not funny; talking about man ignoring human catastrophe and risking human extinction does not come with an after meal mint and a late night bottle of champers. There is no happy ending. There are no winners.

In some sense, perhaps it would be better to outrageously belittle this subject with endlessly gnawing jokes and peppy caricatures making light of things all the way through. But there would be no point. In the end this is a nasty subject, full of death.

Worshipping money is likely a fool's errand; destroying a livable environment definitely is. Who gets over $500 billion in annual subsidies? You? No, it still goes annually to subsidizing fossil fuels and a lot more besides. Our environment has powerful enemies and few friends. If you do not fight, then who will?

All things being equal, the author was not born as a writer nor an environmentalist nor any other small boxed person sitting quietly on society's mantelpiece. He, as many others, is just another human who happens to have tried to care in a world full of doubt and confusion. Bad is good the movies say. Not in this, bad is bad. If I were to try writing a book, I would have preferred to have written about anything else but that, I suppose, is life.

This trilogy is frank; respecting that your time is precious. Three steps, three books. Each requires your action and your choice.

Step 1, this book, answering the Climate Covenant, allows us to understand the consequences of CO_2 and the CO_2 trend. It allows us to create a political mandate for change but only if our combined choice of CO_2 limit dictates that we need to act now.

Step 2, The Votocracy, deals with removing corruption from our political system by giving people the power to make law. We live in a political system corrupted by lobbying and vested interest, creating immoral law not representative of the people's view. We need to fix that. Your vote at the ballot box will decide.

Step 3, The Action, discusses the clean technologies available, introduces a global plan and addresses the clean transition. The choice to enable law to create a clean world will be up to you.

Leaders Showing No Urgency Does Not Mean That There is No Problem

2 INTRODUCING THE CLIMATE COVENANT

As we now know, this whole thing is all about not killing future generations with our action or inaction today and tomorrow.

It seems fairly obvious that that is not a very nice thing to be doing as law-abiding citizens of the planet Earth, or does it? It is real so either we do not realise we are doing it or we do and so are knowingly being complete sh*ts. Isn't it strange, it is fine to kill future people but when it comes to swearing in text, that seems wrong. Or is it that as long as it is truly out of sight, it is out of mind. Either way, the future is our responsibility now.

As a child, I remember being fairly honest with things that mattered and just a little devious with things, like granny's wondrous biscuit mix, that, to me at least, didn't. One afternoon, I arrived home to be confronted by some friends deciding how to kill a furry, harmless caterpillar. I decided action was needed.

So, I rightfully removed them from their bikes before proceeding to chase them hoping for at least a little bloody gore from a few clunks with whatever sharp mass came to hand to thrash against the small podgy human brains now running off into the distance. I had principles, I was in touch with right and wrong and I would not betray it. Killing was wrong, it was simple.

Years later, worrying about right and wrong has become near forgotten. The norm of adulthood seemed to heel to a different drum, changed from things received to needing things, 'must have' things and providing things for others. Obvious morality and happier simplicity seem to have somehow become lost.

We all remember school and most of us remember work. What a dull place offices can be. We are basically back at school, your talentless boss is making another mistake to which you acquiesce, smiling and egging him on in the hope that he'll be sacked and you can slide into an easier position with better pay.

Long gone are the days when you say it as it is; you don't want to give him a chance of upsetting your promising career or your next pay cheque. You are to be too successful.

At some point, it seems that right and wrong simply became incidental or perhaps not truly important but you don't know when that was and probably have no idea exactly why it had happened.

The only thing that really counts is our family, our looks, our wealth, the quality of our friends, the great Waderton Utd and the tabtop to which we cling. Do we even want to care more? We are happy and so why rock the boat? We have no time to worry anyway, and who needs it, worry hurts and no-one wants hurt. Perhaps we are happier not knowing with our TV dancing away in the corner, so what's wrong with that? Well, our inaction is damaging future generations, a lot. It is not if, it is by how much. We may all like to avoid this subject but that does not make any of us right.

We all notice the odd politician, droning on about one policy or another purposefully sending us all asleep so as to avoid the electorate until they wake us up when we are needed by them once in five years with a nice policy carrot to sucker us in, again. Have you ever wondered if what they say bears any relation to what they do? Of course you have and of course it doesn't. Legally it doesn't have to which is another reason why people avoid politics, or more precisely, our current political system.

You probably already know that politicians are the most wrongful beings on the planet quite in contrast to their duty. As this story unfolds, bought policy via lobbying and vested interest, hereby named a lobbyocracy, very much become the problem. Realistically, in order to solve the CO_2 problem, we need to fix our corruptible political system. We will do so but first we must decide when, how and why.

We are assured by the fossil fuel lobby that if the world were to wean itself off fossil fuels then we would end up living in caves. There to settle on hay beds at night by candlelight as the soft whimpering of our tethered cow nestles into her own wondrous dreams cloaked in the warm blanket of night and protected by the magical twinkling of the stars outside. We have all heard disinformation before. Some of us believe it. But it is not true.

There is so much disinformation within this subject that getting the facts straight, without doubt, is a big step to take. The Climate Covenant forces you to do exactly that, then to act upon it.

Originally, the Climate Covenant was to test politicians alone, who would have dutifully answered it, realised their mistake and happily closed down the fossil fuel industry. My mistake. They are informed and they have had 30 years to listen to the science and never have. The time to trust politicians has passed.

The reality is that they know they lie; I just could not bring myself to believe it. My letter was written so carefully that every word and every phrase became the embodiment of a fine tool ready to tease each political neuron into exact alignment. Alas, I then realised that the letter would be about as effective as asking a drug dealer for aspirin. Instead, it is up to us. Our action. Our justification. Our responsibility. Our care. Even our respect for God. If we care, it is in our power to solve. But only if we care.

In truth, what we do for others is far more important than what we do for ourselves, even if we may only realise that too late. We can never have peace unless we care for those to come.

3 THE CLIMATE COVENANT

Before you read the Climate Covenant for the first time, it would be a good idea to first introduce the author and to explain a little more about the beginnings of the Climate Trilogy.

The evidence is that the problem always returns to political power bought by greed. Not new of course, but this time far more difficult to ignore since this time the human cost is horrific.

Power makes law and it is only law that can solve our climate problems. With all power in too few hands, law is easily bought in order to protect industry. Though, with limited media coverage, it is hard to become accustomed to that truth easily. For the author, these books awoke back in 2010 when confronted with by far the most disingenuous piece of journalism I had ever seen. If it had been a joke, it would have been quite funny. It was not.

Having spent some time answering the article in painstaking detail, moments after posting, I was met with what can only be described as the most vile responses from a group of people who were clearly not there for free debate but were there to disinform and abuse. Here was an informed stranger to be quickly silenced. Roiled by the thuggery and bullying dribbling down my screen which faced me quietly at my desk, it was time to ignore the abuse and wait for every last motive to be revealed. From that day to this, it has been very clear that the lobbyist's intervention spreads far beyond their rightful place.

Too many hours later, the bullies were no longer baying for the blood of a cornered quarry, but had tempered and quietened. No longer willing to engage once facts became hard to fight or once lies had been wholly uncovered.

Letting slip the final clue, the name of the lobbying organisation, may have seemed trivial to them but to me was a revelation causing this trilogy to later exist. It was not the reputable newspaper being corrupted that most surprised me but the depths to which the human spirit can sink.

The only recourse we have to tackle rising CO_2 is our vote and the truth. Our vote to tackle the politics and the truth to create action. If you do not use them, then it is likely that inaction will prevail. Hopefully, the truth contained in this trilogy will help you decide what to do. There may be no other voting choice yet; that will change. As of now, this is our best chance to solve this. The Climate Covenant is just a first step, others must follow.

THE CLIMATE COVENANT

Given The Choice of Any Level Of CO_2, Past, Present or Future, Achievable or not, Which Level Of Atmospheric CO_2, To Be Resolved Globally As 'The Climate Covenant', Do You Consider, Due To The Consequences of Which, Must Never Be Breached?

&

In Which Year Do You Consider The Human Race Must Begin Acting On A Plan To Bring Global Man-Made CO_2 Emissions To Zero In Order Not To Breach Your Chosen CO_2 Level?

The consequences include those on the body, the air, sea, land and society such as sea rise, warming, ocean acidification, reef loss and so on causing the loss of health, lives, assets, cities and countries and the increasing risk of food and water loss, war, drought, flood, methane release, other tipping points, extreme weather events and so on played out over the coming years, decades, centuries and millenia ahead, not forgetting the risk of extinction.

Atmospheric CO_2 reaches 1000ppm by 2100 on the existing exponential trend tending to 2.2% annually. This trend is useful in gauging when we must act.

In order to judge the amount of decades that it might take in order to stop all global man-made CO_2 emissions in order to calculate a start year, it may be useful to consider that we have a rising energy demand, in 2013 around 17,000,000MW, the vast majority of products will require replacement or adjustment including power stations, cars, planes, ships, heating systems, military, cooking, factories, houses and so on as well as any process requiring the burning of fossils such as for roads, cement, metals, plastics and so on, including consideration for outstanding technologies such as energy storage.

It is our agreement to Honour The Climate Covenant For all future generations or to damn them & us if not.

GOOD LUCK, TheHuMan

Psychology

Section 2

4 DEATH

Sometimes, in a moment of darkness, we all wonder what death is. Perhaps we wonder when we will meet it and what will happen. We cannot visualise it, we can only guess at what it might be. Is death important? Maybe not, but existence is.

As a young child, I still remember contemplating about what death was, what exactly death was.

I'd be lying there, the light would go out and I would begin attempting to imagine nothingness, an empty black nothing, so that I could visualise and know what death was.

In the darkness, in the silence, I would search the blackness looking for nothing, searching and searching, clearing all evidence of being here from my mind. The world would start closing in, I would start becoming tense and afraid and the world would spin as I forced my mind downwards to an impossible end. Spinning faster and faster until I could only sense confusion. Then, suddenly, it was too much and I would stop, the world would come flooding back to me. I soon grew tired of it always ending up with dizziness, which wasn't helping, but I still remember it clearly to this day.

Years later I do not think of death in the same way. It is no longer a process of discovering its dimensions or characteristics. I know that I will never find that answer in life. However, considering death is still important. It is important because it puts life into context, tempting us to ask questions like 'why are we here?' or 'do we matter?' or perhaps 'does anything we do matter?'.

For me, death is no longer frightening. Though we do not know what follows life, it is possible that death is not the end. Perhaps life has a reason, perhaps it is a lesson or test. Maybe our actions in this life matter. If so, we may dare to go further and entertain the notion that it is possible that our actions really do matter beyond this life and that we may then have to atone, justify or be accountable for them. All religions believe in an afterlife. Of course they may all be wrong but 'Thou preparest a table before me in the presence of mine enemies' puts the idea quite well.

32 | DEATH

This is a grave issue, with great responsibility for each and every one of us. Before making a mistake, it may be wise for each one of us to now decide for ourselves, including consideration for being moral to future generations, our own children, the children of all nations and all other life, whether it is likely that there is a purpose in life, whether one should heed life and whether our actions in life matter or not. Do we have a responsibility here or can we do anything and happily die without consequence? If we do have responsibility, then what do we have responsibility for?

We might recycle or pay some attention but do we really know how much difference we make, what is at stake or how much time is left? Do you know enough?

Does causing death matter? Do future generations matter? Do other species matter? Does life matter? What would we sacrifice for others? What would we sacrifice for strangers? What would we sacrifice at all? When does damaging others matter? Do you fear death?

In truth, this is not about compromising our standard of living for a clean world. A clean world would be better, healthier and cheaper. And save future generations. This is about a choice; one leads to life, the other to death. Without respect for life, this human life will end, one way or another. Tragically so. We were extinct once, we can go back there, and right now, we are.

5 COMPASSION & CONSCIENCE

Perhaps we should, very similarly to considering death, truly try to listen clearly to our consciences, find compassion for others and take full responsibility for our actions, good or bad. I do not exclude myself from the certain implication that we each do not care enough, value oneself too much and others too little. We can change.

We all share emotions, thoughts, instincts and over time we can detach ourselves from each other. Perhaps we could instead come together with a caring voice.

It is a simple point, if we are to solve this problem then we need to act, and to act, we need to care. The caring action will be to stop supporting the burning of fossil fuels in every way that we can.

We design cars, planes, boats and diggers based on fossil fuels.

As soldiers, we are asked to put our lives at risk in order to protect the supply of oil for the fossil fuel industry.

As accountants, we help the fossil fuel industry avoid tax. We work as lawyers, defending companies based on fossil fuels over air pollution, oil spills and fossil fuel emissions. We work as politicians, taking bribes to block clean energy.

We heat our homes with fossil fuels. We walk on the pavement near the road, both made using fossil fuels. We work in the finance industry protecting assets based on fossil fuels.

We fly using fossil fuels, we drive using fossil fuels, we eat using fossil fuels, we sleep using fossil fuels. In current society it is hard not to use fossil fuels.

The typical western person's behaviour releases between 5 and 18 tonnes of CO_2 annually. That is the weight of one car sent into the air each month for each man, woman and child alive.

To produce just one kWh of electricity from fossil fuels releases the CO_2 equivalent weight of one bag of sugar released into our atmosphere. We each use thousands, even tens of thousands, of energy equivalent kWh annually. In other words, a lot. Should we worry?

Today, we are ignoring everything. Every year that passes, how many more people will not live or suffer because of us? There are two paths, clean or dirty. There is only one right choice, yet our world is choosing the wrong one. Are we simply not caring or clever enough? Perhaps we were not designed to beat this issue, perhaps we will become extinct, does that really matter?

It is true that a lot of damage is happening now as a result of burning fossil fuels. Air pollution already kills millions each year. But it is the permanent issues that we are storing up that will cause far worse consequences for the generations to come. What would you say to us, if you were the future life that we sacrificed today? What excuse is enough? Where's the justice?

Perhaps we all need a moment to reflect on our responsibility as humans on this planet, to love ourselves, our fellow man and our surroundings just a little bit more and to have fortitude enough to create a clean world without making any excuses.

The American Indians say that we borrow this land from future generations. I deeply agree. We have never and will never own this world, we each just take care of it for one small moment.

6 CONFORMITY

Pushing my phone onto the bar, I am impressed at my own choice, the latest and the best. Look at what it says about me, I am on the pulse of life, I say smiling. Or could I have been led?

We can convince ourselves that we need anything and maybe it is useful, or useful enough, whether we need it or not. But there is also the aspect of fitting in, belonging and impressing others. Maybe we are less independent of mind than we think we are.

Advertisers often understand conformity well. 'You know what, I am missing out', 'What does my car say about me?', and some milk your vanity, 'You are worth it'.

It is not that we are not worth something; it is that our worth has nothing to do with any product. It is not that we are missing out; it is that we want to fit in. It is not that we should not have the latest; it is that we are not what we own. It is easy to see through these things, but it is hard to act on it. People are easily impressed by money and wealth. Are you? If we are easily enslaved by our inherited God-given weaknesses then perhaps defending against their exploitation is not easy.

We all have that need to fit in, to belong, to feel that we are amongst others that care for us at a very personal level, quite understandably. Realistically though, just being a human-being on Earth is as much conformity as we need. It is as much as we should need. It is as much as we have ever needed.

The issue is not whether fitting in with people around us is good or not; the issue is that we tend to blindly agree with our peers. Worse, we have a tendency to allow our need to fit in be hijacked by strangers, in this case by lobbyists, for a particular purpose which is to make us think as they want us to think.

We are knowingly accepting doubt and lies fed to us deliberately and seem relatively unable to manage ourselves against it. The internet is a powerful tool for that. The fossil fuel lobby has much of the media, TV, web and politicians sewn up. Our only defence is learning the facts, science and evidence, before it is too late. In the end, we are fighting for life rather than death. It is not our differences that make us weak, on the contrary, it is accepting and exploring our differences that make us strong.

The CO2 Trend

Section 3

7 CO2, 2.2% & THE SHADOW

We are human first, not accountants, consultants, builders, teachers or anything else. We are all the same. The same air, the same CO2 and the same responsibility. We may like to think that we are better than others, but it is not true. We are all equal.

CO2 is a gas, we can't see it but it is there and it is not harmless. One extreme harm it can do is similar to the way water is harmful; drink too much and we drown. Increase atmospheric CO2 too much and we drown. Whilst not threatening yet, unlivable CO2 was our early atmosphere and therefore is a possible future.

Something of an example, caused for a totally different reason than the burning of fossil fuels, happened at Lake Nyos in 1986 where thousands drowned in suddenly released CO2. We are hundreds of years, on the current trend, from creating that kind of environment for ourselves everywhere. However, CO2 itself is clearly not harmless even if it is invisible and being ignored.

You probably learnt the basics of life already, that plants munch CO_2 and produce oxygen whilst animals do the reverse. Nature's perfect balance if you will, as it has been for billions of years.

That is until the fossil fuelled world of combustion engine, boiler and factory began life at some time back in the very early 19th century, the 'industrial revolution', which managed to unbalance our planet. We have been ignoring CO_2 problems ever since.

It is the burning of fossil fuels plus the destruction of the planet's lungs by cutting down our forests, known as deforestation, urbanization and desertification, and harming the other lung, the oceans, which is causing CO_2 to rise exponentially, very quickly indeed. We are not blind but we show no care as we walk off the cliff edge smiling as we go.

It is assumed that atmospheric CO_2 will level off or fall when our CO_2 emissions are cut to zero, but what if too much damage has been done? On 'Zero Day', if CO_2 is still rising significantly, what then? We have no global plan to cut our CO2 emissions for the foreseeable future and given population and energy demand increases, far more damage to CO2 sinks will be done. At what point do the risks move from today's reckless to utter insanity?

Right now, and steady for the past few hundred years, the same strong CO2 trend has continued unabated. It is logical to say that the same trend will be our future if we continue with inaction. Simply put, that unwavering trend will likely continue since the factor causing it, our behaviour, will not have changed.

The trend is most accurately described by an exponential rate tending to 2.2% annually or 8-fold per century. By historic judgement, that is an unprecedented, disastrous rate indeed.

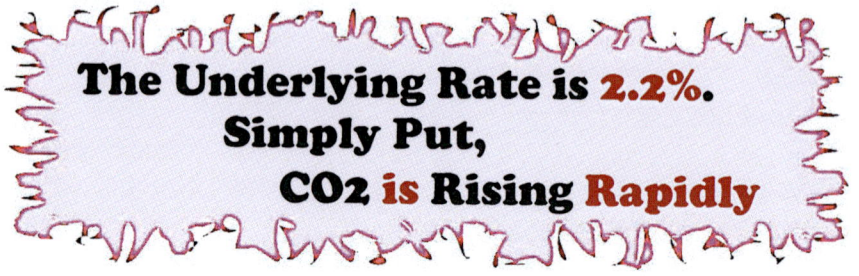

The Underlying Rate is 2.2%.
Simply Put,
CO2 is Rising Rapidly

The current best fit future CO2 trend equation follows. You did like maths, right?! Well, this time we can't avoid it.

$$CO_{2\,(ppm)} \text{ in Year} = 280 + \left(36 \times 1.022^{(\text{Year} - 1959)}\right)$$

As you can see from the equation, there are two terms either side of the addition. The constant 280ppm (parts per million) represents historical CO2 before the industrial revolution began, having not risen above that level for at least a million years.

The variable term on the second line represents the man-made addition. The term 'underlying' is used because, with CO2 at 400ppm, the 280ppm is still dominant representing over two thirds of the whole, whereas the human contribution is currently a much smaller 120ppm; it will not remain that way for long.

The currently larger historic term is dangerous because it is lulling us into not realising how quickly CO2 is in fact rising.

44 | CO2, 2.2% & THE SHADOW

So, with the historic constant larger it creates a 'CO2 Shadow', if you will, which makes it harder to see the true CO2 rise rate. We often ignore our shadow, perhaps this time we should not.

At 400ppm, adding 2.6ppm per year is 2.6/400 or an overall rate of 0.6%. However, our contribution is 120ppm and a rise of 2.6ppm over 120 is 2.2%, the true underlying rate.

By 2100 at 1000ppm, adding 16ppm per year, the overall rate is 1.6%, larger than the overall 0.6% at 400. The human part is now 720ppm, still rising at 2.2%. Therefore, as CO2 rises, the overall rate, 1.6%, is getting nearer to the true rate of 2.2%.

So, as time moves on and the human term gets larger, the overall rate moves closer and closer to the true rate which is 2.2%. Eventually the true rate will be obvious. The true rise rate stays the same at 2.2%, though given the shadow of the constant 280ppm term, that is hard to spot. This is why the CO2 trend is being described as 'tending' to 2.2%, the true underlying rate.

The Constant Term or 'Shadow' Makes the Danger from The 2.2% Rate Harder To See

CO2, 2.2% & THE SHADOW | 45

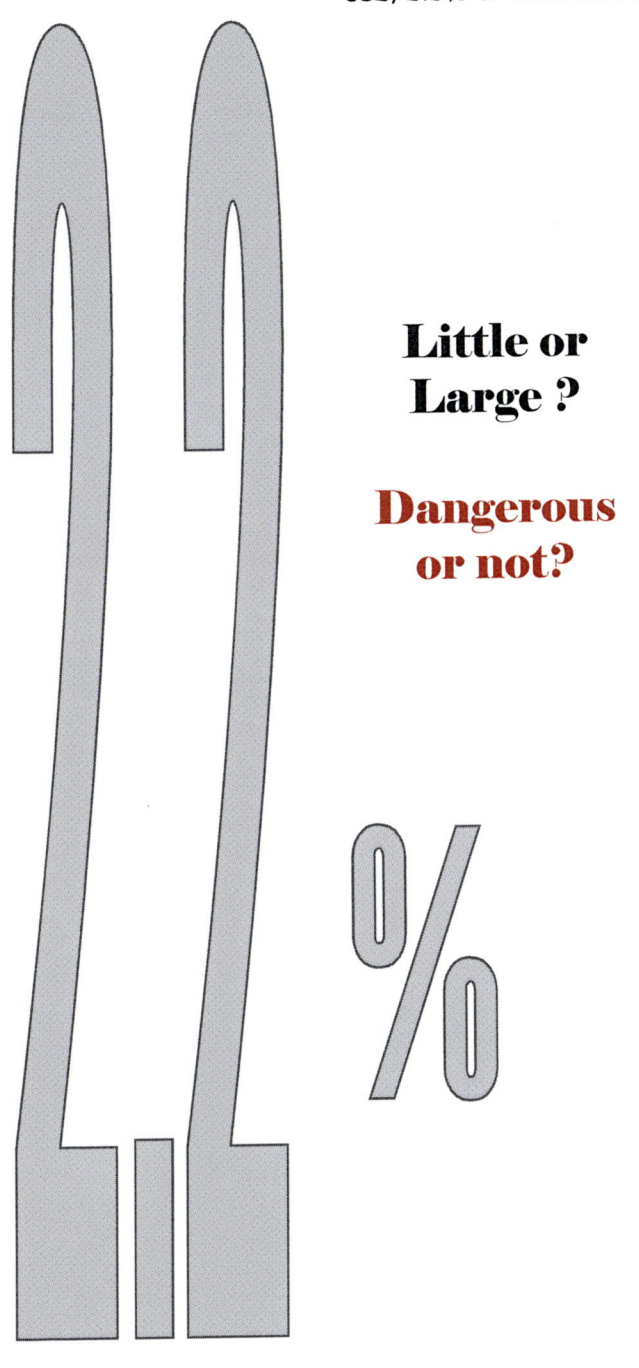

Little or Large ?

Dangerous or not?

2.2%

Our exponential rise is important to understand. It refers to including a variable in the exponent term. So, in our case, the exponent term is $1.022^{(year-1959)}$ where *year* is the variable. What this means is that the amount we add to our total increases each year, as you've seen. Put another way, CO2 rise is accelerating.

For example, back in 1959, when Keeling began keeping accurate records from his Mauna Loa laboratory on Hawaii, the world was adding around 0.7ppm to 0.8ppm annually. This year, 54 years later, we will be adding something approaching 2.6ppm which is over three times more. If we go on another 54 years, then, on our true 2.2% trend, we will be adding near 7ppm or 8ppm every year. As you can see, the addition is increasing.

CO2 is rising rapidly. So eventually, and ever more quickly if we continue to do nothing, we get to all the consequences we know about. The only losing move is to do nothing, as we have decided to do.

This is no joking matter. Things are happening quickly and if we do not know what is coming then we are blind at a time when we need to not be.

The strong trend allows us to extrapolate beyond 1959 into the future. The following chart puts our piece into the entire history of CO_2 on Earth and extrapolates it into the centuries ahead:

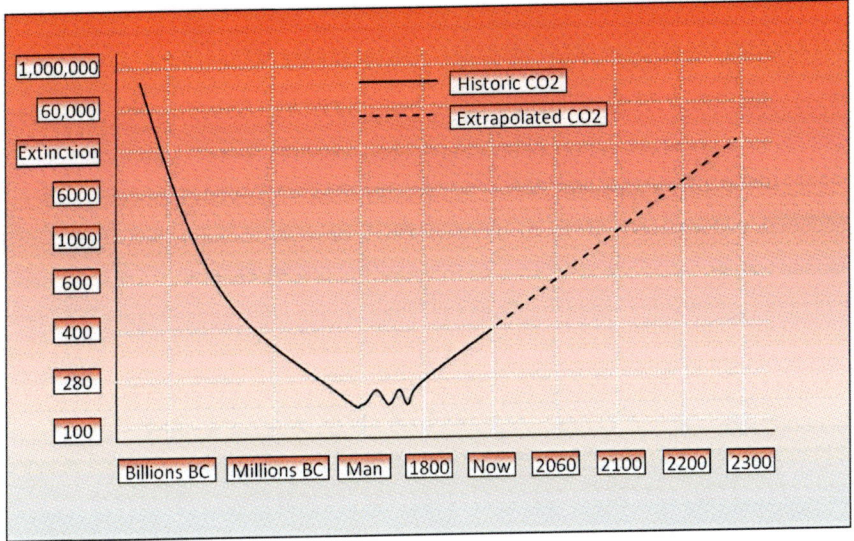

[Chart: Historic CO_2 and Extrapolated CO_2 over time. Y-axis values: 1,000,000; 60,000; Extinction; 6000; 1000; 600; 400; 280; 100. X-axis values: Billions BC, Millions BC, Man, 1800, Now, 2060, 2100, 2200, 2300.]

Whilst the equation is a very good fit for our existing historical data, it is not a world model. The equation does not predict our response to our predicament. However, it is still a powerful guide as to where our CO_2 is likely to go under existing conditions.

For corroboration, the trend extrapolation compares well to the modelled projections from eminent scientific bodies. The IPCC (Intergovernmental Panel on Climate Change) and MIT (Massachusetts Institute of Technology) have modelled a range of scenarios in which the worst cases also near 1000ppm around 2100. The equation may be simple, but it is accurate.

Computer models are often criticised or maligned. The truth is that we have no better way of trying to see what is coming. We do not have a spare planet and spare population on which to test future scenarios. Until we do, we must use what we have.

No-one is heavily publicising detailed predictions beyond 2100 because it seems inconceivable to allow levels beyond 1000ppm to happen. Though that does not mean we won't. On the contrary, many would say it is likely and that is truly scary as you will find out as this book moves through the facts.

The Earth began as a lump of stuff with a horrible atmosphere composed mostly of CO_2 and did not support life anything like we have it today for billions of years. Life began, things grew, CO_2 got eaten and stored over hundreds of millions of years as plants sunk under oceans and fell on land creating fossil fuels such as oil, coal and gas that we are burning in vast quantities without significant thought today.

Oxygen increased and finally an atmosphere allowing humanity the possibility to exist was created. Man has existed for the last few million years, 0.1% of the lifetime of our Earth. Not long. We are not invulnerable. It is possible to return our world to an uninhabitable state and right now, we seem to be trying.

8 WHY UNDERSTAND THE CO2 TREND?

Why? Well, I am sure that you know that this is critical to successfully dealing with this whole issue and so it is worth repeating. If we know when future levels of atmospheric CO2 with existing policies will occur then we can judge when the future consequences of those levels will be and plan accordingly. We would not consider driving blind and nor should we in this.

In any communication or debate about climate consequences the level of CO2 should be established. Without knowing that basic information, there is no basis for debate. Are we talking about today at about 400ppm, fifty years away at over 600ppm or 2100 at near 1000ppm? In any debate about climate change, make sure that you first confirm with the other person what level of CO2 you are both discussing. That question is key.

> **In Debate, Find Out, 'Which Level Of Co2 Do You Mean?'**

When we discuss a football match, do we do it without knowing which teams are playing? Or, do we price a house without knowing which house it is? Of course not, there'd be no point.

Without trying to establish the CO2 level which then defines the consequences, a discussion about the consequences of climate change is rather less meaningful. Find out the CO2 level first.

50 | WHY UNDERSTAND THE CO2 TREND?

The Climate Covenant question requires a moral answer with that CO2 trend in mind. The trend provides the when, you provide the why. Truthfully, there is no answer to the Climate Covenant that can morally justify inaction, economically or environmentally. Those defending the fossil fuel industry know it. That is why the question is set and that is why it is powerful. If confronted yourself by denial, don't hesitate, direct them to the Climate Covenant and let them answer it. If we are not all clear on the need to change, then how is change to happen?

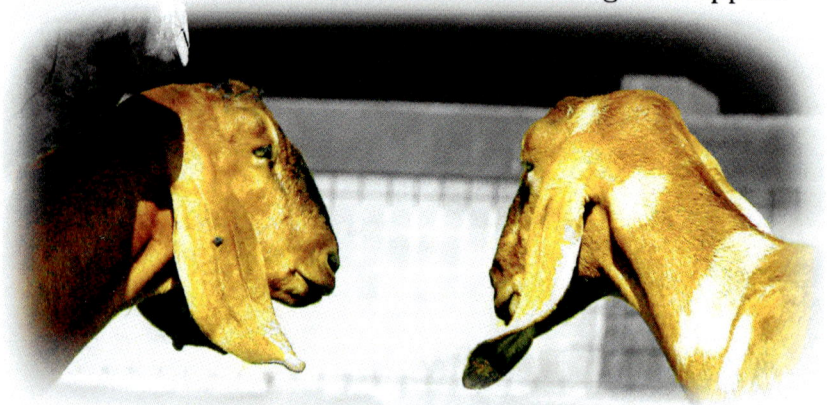

It may be useful to have a basic memory of future trend CO2 levels. So, for our memory, 400ppm now, 600ppm in about 50 years time and 1000ppm at the end of the century.

It is not just words that decide when we need to be tackling this problem; it is nature that does and nature does not lie. The clock is ticking and time waits for no-one. The CO2 data is beyond doubt and can be found at the Mauna Loa lab for us all to read.

Accurately Measured CO2 Is The Only Useful Arbiter Of Our Success Or Our Failure

9 THE BIG LAG

A lag describes the situation where the final result of an action is not immediate. This is critical to understand because the CO2 consequences lag on big planetary, not tiny human, timescales.

There is almost always a lag between every action and reaction. If we drop a glass, there is a lag before it smashes. If we speak, there is a lag before the sound reaches the person we are talking to. Light travels quickly, but it still takes time to arrive.

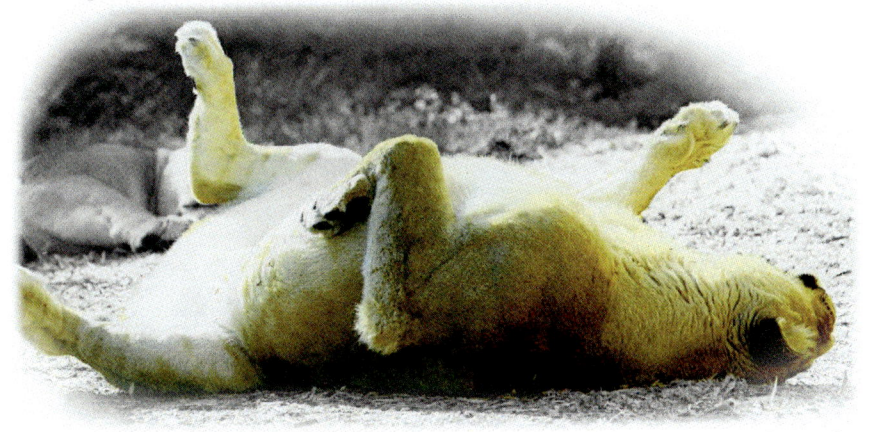

With the environment that basic rule has not changed but the size of the lag has. Historically, man has never had to react to the future damage of actions happening now. That's changed. If we delay action until consequences hit, it will be far too late.

The damage is not happening today and so it is easy to ignore future consequences. This is the danger of the lag, that it is not happening to us and so we can ignore it if we do not realise or do not care. To survive, we need to act with the future in mind.

52 | THE BIG LAG

And by 'Big Lag', it is generally accepted that the consequences of increased CO2 today will play out over the next decades, centuries and millennia ahead. That is incredibly rapidly to our planet, but incredibly slowly to us. Once the lag begins, future generations cannot stop it, no matter how sorrowful we are. We may cause consequences in our lifetimes that will last forever.

Again, to visualise, when an ice cube melts there is a tipping point after which the ice will inevitably melt but it will not do so immediately. We cannot stop the ice melting. We may control money, law, people, companies or countries but we do not control reality; we do not control nature's laws. Ice will melt.

With the planet it is the same though on an incomprehensibly larger scale and way beyond our lifetimes. What we do now affects them not us. It is their lives that we control as well as our own. The danger is that we wait until disaster happens before reacting. What will it take to take responsibility now for them?

For example, if we do wait perhaps another fifty years for consequences to clearly arrive before acting then we will be risking extinction. And it will be future innocents that suffer most, not us, and only the powerful, who are most responsible for this, that may survive with enough money to adapt. However,

Consequences Lag CO2 Rise By Decades, Centuries & Millennia...

against mother nature, none of us may survive.

It is often said that another reason why we should not worry about acting now is that fossil fuel scarcity will force us into acting soon enough. Sadly, that same line has been used for half a century and has never been true. It is disinformation.

It is only action soon that can save lives. The big lag gives us false comfort when the sad truth is that we cannot experience what we have already caused.

The CO2 Consequences

Section 4

10 BREATHING THE AIR

Admittedly, breathing the air may not be something which comes to our minds when considering CO_2 and climate change. But should there be any consequence, however small, arising from higher levels of CO_2 in the air affecting us directly via breathing, then we need to know. Right now, we are risking all levels of CO_2 without any concern at all. That is not right.

We take so much for granted. There is so much that just happens that we would miss if it were not there but are so accustomed to it that we never consider losing it. Air is one of those things. We may not be able to see it, but it is there and we would miss it.

We always remember the crisp spring air full of renewal or the sweet air that drifts through the window as you wind down a narrow lane lined with high hedgerows peering through the gaps to get your first view of the bright blue sea. Then to feel the warm air as it rises up from the beach sand and the strong air as it lifts up a kite to soar away into the sunlit sky amidst the delicate wisps of cloud as if painted just for you. How my spirit awakens, my life enriched by such priceless, simple pleasures. There is nothing in the man-made world that is even close to the importance of our air or any part of this beautiful planet. Should we choose to sacrifice these things? Well, we are.

Perhaps we already have some experience of air not feeling quite so pristine and clean. For example, the not quite so nice air of a crowded lecture theatre or office can become heavy and strangely, despite our best efforts and allowing for the boring nature of our professional world, we become drowsy. An afternoon stretches into a millenia as we try to stay awake.

We are all the same in that we will all react in roughly the same physical way. Bad air may not kill us yet, but it does harm nonetheless. We may accept harm as part of our modern world, but at what point do we decide enough is enough. When our families are affected, our children, all humanity? When we learn what is truly important? Perhaps, by destroying our air?

I always thought that most of the drowsy, headachy thing was just down to being tired and probably having had a few too many buns or pints at lunch time. The desire to have just one more is ever strong and like a wallowing hippo, I seem to relent.

As it turns out, the drowsiness may be more to do with CO_2 than to do with plain chocolate covered morsels of delight. At least my relationship with chocolatey smiles of pleasure can remain an unruly, comforting friend. However, just not eating them may not have been enough to keep me awake. If CO_2 is a problem indoors now then that will only be getting worse.

Of course, we are not only talking about indoor air. What if outside were the same as living in the drowsy office soup of today? What if we sacrifice all our air, wouldn't that matter?

BREATHING THE AIR | 59

One could consider then that the real question is how much CO_2 causes such things? In this case there is no easy answer and not much research. Not that that is particularly surprising since we would need people to volunteer to live for perhaps years, perhaps from birth, in a high CO_2 bubble to find out. So, we have to make the best judgement with the evidence that we have.

Perhaps we should decide that it is better to avoid potential repercussions on incomplete evidence than to wait an unknown period of time for real world certainty and regret it. Having then to accept it even if it has meant real world disaster or real world extinction. We make an idiotic choice, we cannot go back.

From the EPA (US Environmental Protection Agency), air conditioning companies and guidelines for air quality, it is known that air indoors often contains much higher CO_2 than outdoors, in places like offices and schools, due to lack of ventilation, few plants and lots of people breathing.

Outdoors the air is currently near 400ppm and will reach 600ppm in 50 years and 1000ppm near 2100 on our current trend. In other words, outdoor CO_2 levels will reach indoor type levels soon.

Indoors, CO_2 frequently rises by a further 500ppm in office or school environments and can go even higher.

In other confined, controlled or modified environments, such as submarines, the CO_2 levels can be much higher still, though often not for all that long.

In the case of submarines, whilst often cited, they are perhaps not the best guide since all constituents of the air, including CO_2 and oxygen, are artificially managed and the users of that air are not lifelong examples of all age ranges or health levels.

Office research has concluded that high CO_2 has such a physical impact that it affects productivity with impairment starting at 600ppm. Whilst other factors may be contributory including heating, humidity, lighting and odour, it was CO_2 that was identified as the prime cause for the loss of concentration. How close are we to experiencing these things outdoors?

In professional indoor air conditioning terms, it is again possible that problems begin with CO_2 at around 600ppm, as before, with some mild complaints, the volume increasing towards 800ppm and then complaints are more widespread at 1000ppm where the effects include headaches, irritation and breathing issues as the air moves to an officially guided 'non-fresh' state. Things then continue to deteriorate as the level of CO_2 rises further.

The OSHA (US Occupational Safety and Health Administration) 8-hour limit is 5000ppm. Extinction levels are in the 10,000s. The general notion would be that once you experience air at the non-fresh 1000ppm level and have that level brought back to the more normal 400 to 600ppm level then you will never want to experience non-fresh air again. How exact? I do not know.

It is worth repeating again and again, that all of the CO_2 issues are real, are bad and will get worse. The scientific community has been talking about consequences for decades yet still leaders do nothing. How much reason do we need before we change?

Do We Take Too Much For Granted, Including The Air?

11 OCEAN ACIDIFICATION

We all remember litmus paper at school which turned from a drab shade of blue to bright red or green or did almost nothing depending on which oddly named liquid we were instructed to dip it into next. That was all to do with the acidity or alkalinity of that liquid which is known as the pH level.

At school, we may also have learnt that the oceans are also responsible for drawing CO_2 out of the atmosphere thus acting as a major CO_2 sink via processes such as photosynthesis with phytoplankton and the chemical reaction with carbonate creating hydrogen carbonate.

The oceans are incredibly important to sustaining life for many reasons; the CO_2 sink is one, food is another and so on. As atmospheric CO_2 rises, the pH level of the oceans has been and will continue to move lower, in the direction of the acidic end of the spectrum.

As with the air, there is no doubt that this consequence is happening and no doubt that we are causing it. We often make the mistake of talking about this subject as being all about warming, which is missing the point. This subject is about rising CO2 and about all of the consequences of that. One major consequence is indeed warming but that is not the cause nor is it the only consequence, there are many other consequences.

With some of these subjects the facts are certain, but not always. There can be doubt or calculation involved. We tend to assume the best when we should err on the side of caution. We should not always ignore all the warnings and hope that things are ok.

By all means, we could jump into the sea from a cliff without checking the depth of the water below, but it would not be very wise. With CO2, if we don't know for sure, don't jump. The problem is that we don't always know but we are jumping anyway or more accurately, others are pushing us from behind.

This time there is no doubt. The data is recorded and the oceans are acidifying or, more accurately, the oceans are moving to the acidic end of the chart though, for now, are still alkaline. The higher the level of CO2, the more extreme the impact will be.

In terms of numbers, the pH level of the sea is typically around 8.1 today. Any number above 7 being alkaline and below being acidic. Therefore, pH at 8.1 is alkaline. As CO2 rises, the pH number is lowering thus moving towards the acidic end.

To give you an idea of this in reality; we have already lowered pH by 0.1 at 400ppm over pre-industrial levels. At 650ppm that will be 0.3 lower and by 1000ppm, the ocean pH will be lower by perhaps 0.6. Going further, pH will be lower by 0.8 at atmospheric CO2 of around 2000ppm, which will be disastrous.

The effect of changes in the pH of seawater affects marine organism's ability to create their shells and skeletons and also affects their physiological processes. Lots of marine organisms are at risk and particularly the key breeding grounds which are the coral reefs.

It is thought that coral reefs are hundreds of millions of years old. To kill them off in just a few hundred years without caring about what the consequences of that are, is stunning.

OCEAN ACIDIFICATION | 65

So, when will coral reefs all die? Well, many coral reefs are in decline now and have been for many years. When ocean pH recedes to the tipping point of around 7.4 to 7.6 pH, the reefs will die quickly. Reefs will continue to die until a complete global loss would be seen coinciding with about a global 0.6 to 0.8 drop with CO2 levels between 1000ppm and 2000ppm. If we lose the reefs, our breeding grounds, we lose a lot of food.

There are of course other threats to our oceans, ocean warming, over-fishing, pollution and waste dumping leading other scientists to conclude that the coral reefs will be gone by 2050.

According to some reports, one fifth of the coral reefs are already dead.

One last problem, though neither at an existential level nor to do with CO2, is the ever-increasing penetration of tiny plastic particles, named 'microplastic', into our food chain as a result of our Great Pacific Garbage Patches. The microplastic health effects are unknown.

Perhaps man is a tragic disaster, rushing forward into potential oblivion without the merest clue as to why, just like a dazzled yuppie trundling up to the city with no clue as to what is being done in his name, the effect his work is having or how he might miss the enjoyment of working in the much greater outdoors. We have not all been there. We do not all know. We are never as wise as we think. We are not careful. Should we wish to be?

As CO_2 rises and oceans warm, the ability of the oceans to remove CO_2 will reduce then reverse, increasing the acceleration in rise of CO_2 and driving the world into the associated consequences ever more quickly. You can read far more about this if you so wish, it is all out there. Ignorance is defence, but not fairly so.

Oceans Weaken, As A CO_2 Sink, As They Continue To Warm

12 WARMING

It arrives with no fanfare, the ubiquitous warming rears its ugly head, messy and depressing, all at once. We could start with sun spots or the 'hockey stick chart' or the blistering summer heat wave of 1976 or maybe the great gulf stream or what about the greenhouse effect or La Nina and El Nino or Milankovitch.

Or maybe just forget it all and have a pint of ice cold lager having endured the ravages of the desert as did John Mills, Silvia Syms, Harry Andrews and Anthony Quayle in one of the greatest all-time films, 'Ice Cold In Alex'. Isn't nature harsh and wouldn't that ice cold beer have been gratefully received. At times like that, one probably does wonder about the nature of God. Those long wars killed millions in unbelievable tragedy in the name of power. Maybe, as in the film, we should realise that our environment can kill us, all of us, if we care as little as we are.

Most of us like the summer; at least more than winter. So, it is not hard to believe, deep down, that warming might be ok. A bit of warming, how bad could that possibly be?

It is so easy to think fondly of CO_2 with warmer weather, especially in the cold, but truthfully, not rightly so.

Most of us enjoy a game of cricket or tennis or baseball or rugby or football or whatever in the sun on a freshly cut carpet of lush green grass. The sweet smell lingering in our nostrils and the green stains strafing our knees, elbows, fingers and toes. Sweat sheens our skin, the sun bleaches our hair and tans our skin. The day goes on forever. Nature's wonderful ways.

Afterwards to relax, to lie on soft grass looking up at the odd smudge of cloud ever-changing as every incantation of contorted face comes in and out of focus. That amongst the

WARMING | 69

perfect backdrop of deep blue skies criss-crossed by vapour trails left proudly by the metal of excited passengers dreaming of another far-off serenity. We are too little caring of the damage we do. The damage we do has a cost borne by those to come.

Global warming puts all of us at risk. Future generations risk not existing. Unending CO2 rise will take everything that we value away before that reality actually happens. Global warming is one route to damage and human extinction. There are others.

The exact science of the 'greenhouse effect' has been well understood for almost 200 years, first by Joseph Fourier and Claude Pouillet. The science is something to do with GHGs (CO2, methane, nitrous oxide) stopping certain wavelengths and not others. It is easily testable in a laboratory, and so taken as fact.

Simply put, the wavelength that light goes through our atmosphere directly from the Sun is different to that which is reflected back from Earth, whose infrared wavelength is then partially stopped by GHGs, greenhouse gases, trapping heat.

And the effect on the globe has been long expected far before NASA's James Hansen brought the issue to the international community during the 1980s. In 1938, Guy Callendar first proposed the idea that man was contributing to the heating of the planet through rising CO_2 emissions. He was not taken seriously, as is often the way with the conservative nature of the scientific community, but he was right. Now, almost no-one in the scientific community doubts it. It is fact and it is our fault.

It is a worrying indication that now historically quiet scientists are being louder and more enraged than ever before. They feel that they are not being heard. Their facts undermine profit and expose corruption. And because of it the scientists are being attacked, facts distorted, even environmentalists murdered.

Callendar shows atmospheric CO_2 rising linearly to 458ppm by 2200. Whilst he may not have considered or been able to predict exponential CO_2 rise, it is striking to witness that we will be, on our current trend, at 7000ppm not 458ppm in 2200, that is a big difference indeed. Either way the point here is that, regarding global warming, we have been well warned. We have known for almost a century. What will it take for us to listen?

What do our scientists have to do to be heard? They can do no more. It is not a problem with the science. The problem is corruption. Lobbyists buying politicians. Morality overriden by money. Law bought. Lives ignored. Meaningless promises.

G S Callendar & C D Keeling
Thank You

Even if the reason for inaction were economic, which it is not, is there no guilt in costing some lives or all lives? Has life no value? The truth, to science, is clear; we need to act decisively and now. The purpose of this first book is to make sure that you understand it and have a chance to bring your morality to bear.

Why wait? A need for perfect prediction? Apathy? Blocked by greedy, corrupted leaders? Shouldn't we work to avoid the low prediction rather than wait to witness the worst? At least try?

We have been expecting warming and more emissions for a while. The IPCC, in 2007, set out predictions in ranges according to various future policy choices. We are at, or worse than, the worst case CO_2 predictions already. Not good at all.

The current belief is that a doubling of CO_2 will cause the average global surface temperature to rise by about 3°C, called the 'climate sensitivity'. If CO_2 was at 280ppm and 13.8°C across land and sea then a doubling, to 560ppm, will bring average surface temperatures to 16.8°C and at 1120ppm to 19.8°C. That is a major rise and will have dramatic climatic impacts.

Others see sensitivity over 5°C. That means 18.8°C at 560ppm and 23.8°C at 1120ppm. So far, the combined surface temperature has increased by 0.6°C over the 20th century average of 13.9°C, land by 0.9°C and oceans by 0.45°C, and by about 0.8°C since the industrial revolution. The warming will lag the causal CO_2 rise by about three decades. Warming is affected by other factors, including fluctuations in the sun's activity, which cause mini-ice ages and mini-warm ages, though far less significant than man's CO_2 influence today. Recently, the sun's activity has been low.

Of course, the world is 3D. We may need to also consider depth to understand more. It is considered that much of the recent warming may have gone into the deep oceans whose ability to absorb energy and to re-distribute heat is vast and complex. It is considered that one such ocean mechanism, named El Nino, may be responsible for many of the highest upside temperature anomalies through a release of vast amounts of heat from the deep oceans into the atmosphere. La Nina, does the reverse.

The world's governments have agreed to keep the world to less than a 2°C rise. That is keeping CO2 under as little as 400ppm but certainly under 450ppm by consensus. 450ppm, being 17 years away, means aggressive action now. Their action, by contrast, is to plan to do nothing before 2020. In short, politicians lie.

> **Governments Agree 2°C Limit Yet They Still Block Action. Do You Still Believe That They Are The Solution Or Were They Always The Problem?**

We may ask why the figure of 2°C has been chosen. That is the figure that ministers say, presumably on political assessment of the science, will provide us a chance of avoiding catastrophic climate change such as extinction caused by tipping points. Are they right? Well, long after it being avoidable, it is true to say that some leading scientists strongly disagree with the limit of 2°C, they say it should be 1°C or 1.5°C instead. They are still being ignored. Rather we should consider ourselves lucky that they give us warnings at all, often at great personal cost. Instead our politicians see them as an inconvenience, as they do our lives.

Scientists are conservative, for very good reason; their livelihood depends on accuracy. They may hold belief but, with regard to science, if it cannot be peer proven then it can't be publicised. The media, by complete contrast, live under a different mantra; money. They are able to avoid truth similarly to our politicians. Backed by lobbyists, if the facts threaten profits then a scientist's every word is assaulted. Lies born of fabricated doubt become public truth. The result is fabricated justification for inaction.

We cannot blame scientists. It is not a scientist's job to be a politician or a politician's job to be a scientist. Politicians should listen and be moral, instead they bend truth for greed whereas scientists strive to uncover truth for the public good. 350ppm was our clear safety point, that is long gone. We are being told not to breach 450ppm, though we plan to. Where does it end?

Another critical aspect of warming to understand is that it can lead to our extinction through the crossing of tipping points such as methane release. Extra CO2 rise can also follow warming and ocean currents, such as the gulfstream, slowing down can cause local extremes. Humidity due to warming is increasing, rapidly worsening flooding via heavier rainfall also limiting the viability of us living in some geographic areas at all by limiting the human body's ability to cool via sweat. The data can be seen at NOAA (US National Oceanic and Atmospheric Administration).

13 SEA RISE

One of my favourite things is going to the beach and enjoying the sea. Soon enough, that may not be quite as easy as it was. How strange that would be; how we'd miss easily lounging on golden sands drinking cocktails whilst applying another layer of lush smelling coconut cream. Drowsing happily with the breeze caressing every curve of our bronzed bodies as we watch the gentle waves break on the shore in comforting sound alongside the beautiful blue, sparkling sea. You may say that there will still be beaches, 70m later and thousands of years on. If we are still here, at best, yes, some may be. But why accept this, just for a few rich men, surely nature is more valuable to us than this.

Sea rise comes about as a result of global warming. As the land, water and air heat up, the water expands and ice melts. Sea ice, such as in the Arctic, adds little to sea rise but the land-based ones, Greenland and Antarctica, can add incredible amounts.

76 | SEA RISE

The Greenland ice sheet is said to embody a possible 7m sea rise and Antarctica a further 60m. Including glaciers, the possible rise is around 70m. Ice cores and meteor craters give us an insight into historic sea rise over millions of years. Projections using this evidence give us a range of possible future scenarios.

The poles are warming three or four times faster than the global average with Antarctica warming 2.8°C since 1958.

It is believed that a rise in temperature of below 4°C locally, or near 1°C globally, will cause Greenland's ice sheet to proceed to melt. At 2°C and 450ppm, future sea rise will get to 7m as that ice sheet melts. Others see evidence for over 5 times that.

Thermal expansion of water has a small climate sensitivity adding 1.25m per CO_2 doubling.

The full 70m is a real risk and the lag for the full melt is seen to extend into millennia, it may be quicker, no-one can be certain.

This is another area where disinformation is rife but the facts cannot be disputed. It is clear that the Arctic is melting quickly, aerial photos confirm this certainty, as do other surveys. Though the Arctic melt will not cause sea rise, Greenland and Antarctica will. As the melt happens over centuries, it is easy to cast doubt no matter how overwhelming the science and evidence is.

Regarding the Antarctic, as you know, it has a local warming trend much higher than the global surface rise by around a factor of 3 times. A local rise in temperature of 10°C is believed to cause melting to begin with a rise of 20°C believed to be enough to cause a complete melt of the entire ice sheet, equivalent to a 3°C to 7°C global warming level, possibly starting with CO2 as low as 400ppm but certainly significant melt by 600ppm.

As mentioned earlier, there's evidence linking sea rise with lower levels of CO2. Some believe that at 400ppm we are already in for a 25m rise. A Siberian meteor crater suggests higher; 40m at near today's CO2 levels. Analysis of the ice cover of Antarctica going back 34 million years, estimates that the tipping point may be 1000ppm for a complete ice melt and 200ft sea rise.

In all of this, as we know, science tries to provide a conservative view, a minimum if you will. In a way, that is fair enough and would be fine for predicting anything where undershoot does not take lives. Sadly, under-estimates in this risks extra devastation and extra lives lost. Since it is impossible to know for sure, would we be best served by heeding worse-case scientific projections rather than the average or the best? Perhaps our reaction needs to be decisive not hopeless, certain not dismissive, careful not careless. There is no safe option. It is a choice between future disaster or future extinction. Do we still want to do nothing?

The flooding due to sea rise is likely to be catastrophic and play out over centuries and millennia to come. No country will avoid the impacts of that flooding and many coastal countries and cities will cease to be. Many unknowingly are already damned.

Low-lying counties or states plus countries like the Netherlands and waterfront cities as London, New York or Miami will likely be devastated by CO_2 levels that we cannot now avoid because of political inaction. Is it right to continue to ignore CO_2 rise, just waiting to see what is destroyed next? Well, we are.

At 70m all countries will have ceased to function as they do now falling back into a much less stable sea world having lost much of their former infrastructure against local problems and global strife. Small sea rise is bad but massive sea rise is a nightmare.

Put into perspective, in 2070, given a small amount of sea rise, an OECD (Organisation for Economic Co-operation and Development) study calculated that by 2070, considering just port cities, $35 trillion in assets would be put at risk by flooding. And that is just the bare beginning. That will not be insurable, so who pays the bill?

Economically, there is no contest. It is cheaper to have a clean world than to pay for the fossil fuelled consequences. Is it time to wake up?

We can find sea rise maps across the internet. You can happily destroy your town without getting any toes wet. Soon though it will be for real. Losing everything is not easy, it is easier to act.

14 MORE FOSSIL FUELLED CONSEQUENCES

The first four consequences are consequences that cannot be addressed through any form of adaption that does or is likely to exist. They are a relatively permanent gift to future generations. Because of our choice not to mitigate our CO2, the world will not be the same again no matter what we do or what our successors do. There are other consequences. Some we suffer now. Others will cease once we stop burning fossil fuels and some will not.

Sadly, many of these smaller consequences affecting us daily are still rarely linked to the cause, CO2, or the problem, systematic lobbying by the fossil fuel industry of our politicians, by those on TV or other media.

We will now mention a few of the other fossil fuelled issues to help make things clear. That is, if you are still unsure if suffering the consequences of burning fossil fuels is worthwhile or not.

MORE FOSSIL FUELLED CONSEQUENCES | 81

Fossil fuels are known to leave tiny particles in the air made of metal and other by-products of combustion. This is called 'air pollution'. This particulate matter, such as PM2.5, is literally breathed in and passed via the lungs into the bloodstream where it finds its way into damaging every part of our bodies.

The air pollution is said to shorten western lives by a year and up to over 5 years in parts of China. In some cities, it is the equivalent of smoking one packet per day, with advice being to not go outdoors. It is said to account for over 2% of global deaths or millions each year through effects such as asthma, cancer and heart disease. Cities are worst affected, as are infants. A silent killer, but a killer nonetheless.

In terms of the land, as the air and land temperature rises, deserts expand. It is called 'desertification'. The problem is widespread and worsening, it already affects many countries in the world today and one of its results is a large loss of agricultural production and food.

The stress on food supply affects society as prices rise. Food shortages can and do cause social unrest, rioting, and, in less stable areas, are used to control population or even cause war.

82 | MORE FOSSIL FUELLED CONSEQUENCES

This phenomenon is happening now. As the population of the world increases and the land suffers from drought, fire, flood, less fresh water and desertification, the pressure on food supply increases and people fight in the streets for basic provisions.

In terms of water, only 3% of water on Earth is drinkable. As our population grows and the need for fresh water rises, sources will be falling. As snow caps and glaciers shrink, so will many rivers and lakes. Groundwater will become salinized as seas rise. In a little over a decade, almost two billion people will be in physical 'water scarcity', meaning they do not have local access to enough drinking water. Desalination, turning salt water into fresh, is one possible solution but is energy intensive, potentially making the cause, our CO_2 emissions, even worse.

MORE FOSSIL FUELLED CONSEQUENCES | 83

As things progress and the world moves into a higher CO2 state, other dangers increase and other factors emerge.

One of those is the release of methane which has far stronger greenhouse gas characteristics than CO2 but lasts for less time in the atmosphere.

As the world warms, so-called 'tipping points' may be reached whereby a certain key temperature has been met, ice caps melt, permafrost becomes exposed and suddenly methane starts entering the atmosphere in large quantities.

That causes an acceleration in warming, which itself causes more methane release and so on. This is called 'positive feedback' where the consequence causes more of the same consequence. It is believed that runaway tipping points may have played a role in previous extinction events, some culminating in raised levels of CO2 or perhaps even in lethal hydrogen sulphide exposure.

84 | MORE FOSSIL FUELLED CONSEQUENCES

There will be an impact on our weather, as we are starting to witness already. It is not the case that just one piece of weather is directly affected by elevated CO2; it is that every piece of weather is affected but no-one can say by exactly how much. The truth is that we will see more extreme weather as a logical by-product of more energy existing in our climate system, increased humidity, sea rise and so on. Simply put, as CO2 rises, the level of extreme weather will worsen for centuries ahead.

One unseemly factor to consider is the effect on asset values and how that will be handled by the financial sector. The likelihood is that the industry will look after itself and get set to gain from an inevitable shift, one way or another, in fossil fuel related assets. The point being that we either plan for this or we leave it to take a natural, likely distasteful, course.

Finance is another industry with a powerful lobby and poor moral record.

As the seas rise, there will be countries that will cease to be and that will lead to forced migration. Who will pick up the bill?

As the supply of food and water decreases and population increases, countries will be less capable, then less willing to accept ever greater numbers of refugees. The quality of life many enjoy now will be a thing of the past.

In a perfect world, much of this upheaval could be limited. Though that is almost certain to be wishful thinking as it depends on the same missing understanding and care that has caused the issue to begin with. If we do not care now, why would we then?

In not too long, it would not be at all surprising to see an increase in armed conflict and war as compassion wears thin. Should the world have waited too long to deal with CO2 and forewarned, foreseen tipping points begin, events may unfold too quickly and disasters be too big for the international community to cope.

Again, if handled compassionately, the world has the option of coming together in this crucial time of need and putting aside petty greivances, egos and vanities. Great strife may force the human race out of its tribal past but only if we can show more care. If future generations had our choice, they would tell us to grow up and to begin listening to what the facts have to say.

Answering The Climate Covenant

Section 5

15 CHOOSING YOUR CLIMATE COVENANT

This is where you help decide our future. This is where you use your knowledge, throw away your trust in hope and begin to say goodbye to your apathy. How fair will your answer be?

Your answer is to decide if, why and when we need to act. This is where you shoulder some responsibility or ignore it. No big action yet, no money asked, no blame, no jokes, just a very important decision and the responsibility that goes with it.

It is certain that the world is not acting, it is certain that we are making a mistake; it is not certain if you agree or what you want to happen. If we do not make change happen, then the world will slide into increasing disaster, whether we like it or not.

With your help, we will fix all that. First we create an indisputable reason, next we deal with our leadership and finally we act.

CHOOSING YOUR CLIMATE COVENANT | 89

The Covenant table at the end of this chapter is designed to help you take your decision, that is to decide the CO2 level beyond which the consequences are too bad for you to want to pass on to future generations, if you have not already made that choice.

The table shows the CO2 levels that you need to choose between in order to answer the Covenant. Each is differentiated by key levels and events within the four non-adaptable consequences; the air, the ocean acidification, the warming and the sea rise.

In the case of rising levels such as with ocean pH, warming, sea rise, flooding and lives lost or affected, when you find a figure which you do not accept then the corresponding CO2 level is too high. Your Covenant level will be one of the lower CO2 levels shown above.

The events, such as air damage, reef damage or tipping points, are shown in yellow from the CO2 level at which they start. If one such event starts in your row then you have not accepted it by choosing that row. However, if you do not accept it then you cannot move to a higher CO2 level. You must accept all the consequences in the row you choose or you must choose a lower CO2 level from one of the rows above.

90 | CHOOSING YOUR CLIMATE COVENANT

So, if you do not accept the risk of extinction through tipping points then you must choose no higher than 400ppm. If you do not accept flooding to cities such as Guangzhou, London or Miami, then you must move from 400ppm to 350ppm or lower.

Be clear, your CO2 choice states which consequences you are willing to accept on behalf of humanity not which CO2 levels it is still possible to avoid. We are establishing whether people want to deal with CO2 now when politicians do not and are not. If that answer is yes, then the next step moves to the politics.

The following describes what each column in the table means:

CO2 Level

This contains the CO2 level choices for your Covenant answer.

Air

This contains a description of how our experience of breathing the air might change with respect to each corresponding atmospheric CO2 level.

Ocean Acidification

This contains two columns. The first, 'pH', shows the range of acidities of the oceans for the level of CO2 in the first column. The second, 'Effect', shows the impact that those pH levels will have on our reefs which in turn affects our food supply.

Warming

This contains five columns. The first two, 'IPCC & Consensus', show the warming expected at a climate sensitivity of 3°C and then the resulting sea rise associated with that warming.

The next two under 'Other Evidence', show warming associated with a 5°C climate sensitivity corresponding to other sources of evidence and opinon and the corresponding sea rise related to that evidence and to that climate sensitivity.

The last column, 'Possible Warming Lag Year', gives an idea of the lag involved in the warming by providing a lag year. This is not a prediction, just an idea of the possible kinds of lag involved.

Sea Rise

This contains four columns. The first, 'Specific Rise' combines the range of sea rise taken from the two sea rise columns in the prior 'Warming' section.

The next, 'Possible Lag Year' represents a rough guide to the kind of lag expected with regard to the sea rise represented in the first column.

The last two columns named 'Catastrophic Flooding' list some examples of countries and cities which will suffer levels of permanent flooding at the sea rise shown which should be considered catastrophic by size of geographic or asset loss.

Lives Seriously Affected & Lost
(per future global population)

This column represents the numbers per generation who are likely to be seriously affected, or never live, either directly or indirectly as a result of the four main consequences. In other words, those affected by issues including water shortage, land loss, food supply, disease, migration, permanent and temporary flooding, fire, heat waves, war, humidity, the air and so on.

The numbers affected by the temporary effects of burning fossil fuels, such as air pollution, will worsen; don't forget that either.

There is going to be some doubt, so don't complain about it, we do the best we can. The choice is between bad or very bad, not a choice between good and bad; there is no doubt in that. Nor

CHOOSING YOUR CLIMATE COVENANT | 93

is there doubt in the fact that a choice has to be made. Nor in the fact that it is us that has to make it. This bit is hard, we are choosing the future on behalf of future generations. We have a rare opportunity to do something truly meaningful and truly wonderful for those to come. By our actions now, we can save many millions, even billions of lives. Or condemn them if not.

This is your action in this first step and the thing to commit to memory; your start year and your Climate Covenant CO_2 level decided from the consequences associated with it; take your time and get it right.

Remember as much as you can; that is your responsibility from this moment onwards. We can no longer plead ignorance or hide behind any prior innocence. The tables and recap follow, read them all carefully.

CO2 Level (ppm)	Air	Ocean Acidification		Warming						Sea Rise Consequences			Lives Seriously Affected (per future global population)
		pH	Effect	IPCC & Consensus		Other Evidence		Possible Warming Lag year	Specific Rise (m)	Possible Lag year	Catastrophic Flooding		
				+°C	Resulting Sea Rise (m)	+°C	Resulting Sea Rise (m)				Example Countries	Example Cities	
280	Fresh Air Outdoors	8.2	None	0	0	0	0	1800	0 to 7	2300	None	None	None
350			Increasing Reef Damage & Reef Death	0.75	2	1.25	7	2019			Island States, Tuvalu Netherlands, Bangladesh, Egypt	Amsterdam Rotterdam Alexandria	100,000s to 10s Millions
400		8.1		1.25	4	2	25	2044	4 to 30	2500	UK, Japan, India Thailand, Denmark Vietnam, Greece Sweden, Estonia	London, Miami New York, Norfolk New Orleans Norfolk, Toronto Ho Chi Minh City	Millions to 100s Millions
450	Possible Air Impacts Begin Headaches Drowsiness Asthma			2	7	3	30	2060					
600		7.9		3.25	10	5.5	40	2089	10 to 40	2700	US, China, Brazil Germany, Belgium Kazakhstan, Iraq Finland, Australia New Zealand Russia, Canada France, Italy	Osaka, Bangkok Hong Kong, Tianjin Dakar, Shanghai Mumbai Guangzhou Nagoya, Venice	10s Millions to Billions
800		7.7		4.25	16	7		2112	16 to 70	3100			100s Millions to Billions
1,000	Fresh Air Lost	7.6		5.5	35	9	70	2127			All Countries	Almost All Cities	
2,000		7.4	All Reefs Dead	8.5	70	14		2167	70				Billions
5,000	OSHA 8 hr limit	7		12.5		20.5		2213					
20,000+						Extinction							

Increasing Risk of Runaway Tipping Points

Don't Forget: Lives lost & harmed by damage to nature, disease, loss of species, crop damage, water shortage, food shortage, extreme weather, tornadoes, fire, flooding, drought, heat, humidity, shortened longevity, deaths from pollution, war, murder, oil spills, corruption, greed, corporate profiteering, mass migration, methane, tipping points and so on; *they are all fossil fuel consequences*

17 WHEN SHOULD WE START?

In any rational sense, the world is facing by far the worst consequences in human history including facing the risk of its own extinction. World wars are probably the closest comparison and even those will be dwarfed by this. Our reaction has been to block clean solutions at the political level and to largely ignore the problem at the private level. The result of that has been near total inaction with nothing more than headlines and words to attest the opposite.

That is clearly wrong if easily explainable but it is not forgivable when so many lives will be affected and lost. We are taught to forgive, but how much forgiveness can there truly be?

That may sound a bit weird; we like to think that we have started to deal with our CO_2 emissions with Kyoto and so on. Unfortunately, as we can see from the CO_2 trend, nature disagrees with us and we simply can't argue otherwise. The CO2 data doesn't lie even if we do.

So, by 'starting', we do not mean hoping to address part of our electricity supply which is only part of our energy supply and part of our problem which includes cars, planes, boats, boilers, power stations, factories, military, cookers and so on, at some far off point in the future which we then do not honour anyway. Our current political stance is to do nothing, cause damage and be proud about it. No more.

'Starting' means tackling the whole problem, all our CO2, with an unbreakable plan to cut man-made CO2 emissions to zero, without omission, with certainty, done absolutely transparently. It is not a difficult point, either we are addressing the problem or we are not.

Right now, we are not and have never been.

Our global fossil fuel energy usage stands at a lofty renewable equivalent of perhaps 80,000,000MW today. A medium solar farm or wind turbine would be a few MW. We have a lot to do.

Some of us, every year or two, try again to get fit. It's spring and the sun is starting to make an effort as we look at ourselves in the mirror hoping for one thing and finding something different.

After a quick calculation, confirmed by an involuntary shiver, it is realised that we only have a few weeks left before we unveil ourselves on some distant exotic shore in the hope of enjoying a well earned pampering, which only the delights of that far-off destination can surely provide.

With a rush of enthusiasm and a vague notion that others are doing better than us but that we can catch up, the gym becomes our sanctuary. A place to improve ourselves. So, muscles are hurt, hearts pumped, sweat surrendered, fat wobbled and for a while the improvement sought is close. But it is a lot of effort.

Surprisingly quickly, it all wears off, the membership lapses and we are consoled that we are not so bad; at least we tried and almost succeeded. We can always repeat it next year, and anyway, it is all now too late. Simply put, we had no will.

That is about what happens with the CO_2 issue except that so far we have not even made it to the gym. We are not even sure that we need to do anything. That must change.

The table shown later breaks down the scenarios for us into 50, 75 and 100 year transformations from a world dependent on fossil fuels to one where anything dependent on burning fossil fuels has been replaced by something clean.

To be clear, we need a true clean world meaning addressing all processes or products producing CO_2 emissions. In short, anything based in any way on burning fossil fuels.

From the moment that the world tries to 'start', on some future date, there will be a delay between that and having the global agreements, infrastructure and mechanisms in place to enable the transition to occur. All of that will take a great deal of time.

Then the world will need to begin the financing and research phase where solutions are developed. Some of that exists now but there is a lot more to do; power, heaters, energy storage, roads, cars, transport, military, boats, planes, factories, cement, steel and so on. After that, the world will need to scale up by large amounts in order to meet the global scale of transition.

That global agreement, infrastructure, financing, research and ramp-up phase could perhaps take between 10 and 20 years given the fact that the world has spent 75 years without being able to do anything useful, let alone deal with the considerable logistics and problems that real action would involve.

Perhaps the shortest possible time for orderly transition is 50 years with the longest perhaps being 100 years. 3% growth makes anything longer illogical and assets make anything shorter unlikely. Except, that is, for a war-like response to a disastrous climate event though, if we wait that long, no-one'll likely survive at all.

Of course, a rational mind would have acted long ago given the scale and variety of consequences coming towards us and the level of unforeseen risk involved. Given that there are no technological nor economic barriers, and have not been for years, something is very wrong. That main problem is undoubtedly corruption borne of politicians being bribed by the fossil fuel lobby to create inaction via law. This book deals with the disinformation, the next with the political corruption.

Is the world right to risk human extinction before tackling CO2? Is it right that lobbying rules? Is it right that cities and countries will be lost? Is it right to damage our air? To destroy the oceans? To create war through food and water shortage? Ignore future lives? That is what we are doing, so maybe we need change.

The table asks you to choose between a fair 100, a fast 75 and a tough 50 year plan depending on what you believe likely, all with 15 year agreement and development periods after which a smooth global transition to zero CO2 emissions is assumed. The equivalent CO2 peak is expressed as being on the current trend halfway between the 15 year period and the plan end.

WHEN SHOULD WE START? | 101

As an example, the peak CO2 level after a 50 year plan will be found on the trend line 33 years after the plan start date. For the other plans, 75 and 100, it is found at 45 and 58 respectively.

Obviously acting soon is important. Less obvious is that acting sooner is relatively more important than the time taken to finish. I will explain.

If it were 1972 and you wished not to breach 450ppm, then you can happily embark on the 100 year plan.

However, if you do not, and wait 25 years later until 1997, then you have to choose the 50 year plan to acheive the same 450ppm. So, a 25 year delay has given us 50 years less to meet the same CO2 level. Every extra year of delay cost two in delivery. The reason is that the cause, CO2, is rising exponentially which means our problems get worse ever quicker as we wait.

Starting Sooner Is Relatively More Important Than The Length of Plan Chosen

WHEN SHOULD WE START?

In the transition table below, by finding your Climate Covenant level of CO2 in the first column, you can then follow the row across to find the start year for the transition plan of your choice. You need that start year to complete your contribution to the Climate Covenant. I wonder, what will your answer be?

Within the table below, some start years are already in the past. If your start year is in the past, as it is very likely to be, then you have made the critical decision, that waiting time has run out. If waiting time has run out, then it is time to act, which is the point of this step. The Covenant is a combination of all answers. Your answer is your morality, the level that you do not want to be breached. That answer is valid regardless of whether it is breached now, is breached soon or is yet to be breached.

| Your CO2 Level | Transition Plan ||||||
| | 100 Year (fair) || 75 Year (faster) || 50 Year (tough) ||
	Start	End	Start	End	Start	End
280	1800	1800	1800	1800	1800	1800
350	1931	2031	1944	2019	1956	2006
400	1956	2056	1969	2044	1981	2031
450	1972	2072	1985	2060	1997	2047
600	2001	2101	2014	2089	2026	2076
800	2024	2124	2037	2112	2049	2099
1,000	2039	2139	2052	2127	2064	2114
2,000	2079	2179	2092	2167	2104	2154
5,000	2125	2225	2138	2213	2150	2200
20,000+	2191	2291	2204	2279	2216	2266

18 MAKING UP YOUR MIND WITH A QUICK RECAP

You're tired, and rightly so, perhaps a weary journey full of unsavoury surprises. Or perhaps, better, an informative read entwined with moments of vague, childish nostalgia leading to an interesting hypothesis expressing a hugely important and highly worrying truth; that time has run out and still the world does nothing. That is, nothing that the real world has noticed.

Without nagging or selling, without causing embarrassment or unduly pressing you further, we shall, with the genuine heartfelt promise that this whole thing is to help you make the best decision for your own conscience on behalf of future generations, take you through a recap of the probable reality expressed within these pages. Directly, harshly, quickly and without any compromise to nicety or to saving vanity. We need to arrive at a caring, considered and informed answer. The Climate Covenant question is set out below, read it carefully.

**Given The Choice of Any Level Of CO_2,
Past, Present or Future, Achievable or not,
Which Level Of Atmospheric CO_2,
To Be Resolved Globally As 'The Climate Covenant',
Do You Consider, Due To The Consequences of
Which, Must Never Be Breached?
&
In Which Year Do You Consider The
Human Race Must Begin Acting On A Plan To Bring
Global Man-Made CO_2 Emissions To Zero In Order
Not To Breach Your Chosen CO_2 Level?**

It is our agreement to Honour The Climate Covenant For all future generations or to damn them & us if not.

104 | MAKING UP YOUR MIND WITH A QUICK RECAP

It begins with atmospheric CO2 at 280ppm before the industrial revolution began. That is no consquences, no risk, no sea rise, no warming, no ocean acidification, no deaths and no damage. You are forgiven for choosing that level, it does make sense.

At 350ppm the sea reefs begin to die. Seas risk rising up to 7m and we expect around 1°C of global warming, give or take. Even with consensus sea rise at 2m, countries and cities will be damaged. Port cities alone are said to risk $35 trillion by 2070. If you don't want more, you want to save lower-lying countries and cities and you don't want the coral reefs dying or millions seriously affected, then this is your level. With the faster plan, your start year choice to tackle global CO2 emissions is 1944. Do you agree with the scientists, 350ppm is the pragmatic limit?

Has

With CO2 rising further to 400ppm, global warming now risks 2°C with the consensus at over 1°C. The seas rise to over 4m and there is a risk of a catastrophic 25m, perhaps half a millenia away. That threatens all major countries in some way with disastrous flood damage. Coastal cities will be damaged and at worst, decimated. Reef damage worsens further. Millions are affected with the chance of that rising to hundreds of millions.

We are, for the last time, able to avoid catastrophe for humanity by not breaching 2°C and so not risking tipping points which, as you know, can create positive feedbacks such that we cannot then stop, no matter how much we object, leading us to extinction.

MAKING UP YOUR MIND WITH A QUICK RECAP | 105

If you choose 400ppm and the tough, fastest transition, the world needs to start its plan to replace power stations, cars, planes, boilers, factories and so on by 1981. Is this enough damage and risk for you; a world sinking alarmingly below the waves? Disaster certain, millions of lives affected or lost?

At 450ppm, 17 years away, the consensus is 2°C and risking 3°C. We are now risking extinction. The world's politicians agreed to go no further than this but they are too often corrupted by lobbying and self-interest to honour their rhetoric. Bluntly put, they block action in support of corporate sponsors. If the world is to avoid disaster, it must deal with political corruption first.

You should now begin to realise what we are up against, what is at stake and that should send an unmistakable shiver through your soul. This is not Hollywood fiction ending at the credits to be replaced with pizza and pop; this is reality. The fair plan begins in 1972. If you like coastal loss but don't like extinction risk caused by runaway tipping points, then is this your level?

Waiting Time

The next level is 600ppm. This is quite a jump with warming well over 3°C and perhaps nearing 6°C, air becoming problematic for possibly the first time outdoors, indoor air correspondingly worse, perhaps half-way to losing all sea reefs, sea rise over 10 metres and possibly many multiples of that. All major countries affected. People seriously affected rising to catastrophic levels.

600ppm comes with disastrous problems guaranteed. The fair plan starts in 2001. This is not a sane level to choose. If you like catastrophe with massive sea rise, large warming, dying reefs, untold numbers affected and further increasing extinction risks but don't like risking the air, then is disaster your chosen level?

Let's be absolutely clear. The point of this first book is for you to decide if waiting time has run out or not. It does not matter if the level chosen has been breached already or not. You choose the level with all consequences you accept but no more. If you choose catastrophe then so be it, you will have to bear the responsibility of that choice even if you do not have to witness the real consequences of it. Your answer contributes to one answer, the Climate Covenant. That one crucial voice states, unequivocally, if we believe waiting time has run out or not.

Run

From here, all risks are in play and they all simply worsen until we act or are forced to become extinct. At 800ppm, breathing is likely affected, we get at least 16m of sea rise and risk 70m. At 1,000ppm, 6°C of warming but risking 9°C. A billion are affected. Society is near certain to collapse. All because we did not act.

Tipping points are likely causing positive feedbacks meaning that we can no longer prevent further warming and further CO2 rise. Nothing may now prevent it. From here we are in the hands of God. Perhaps we might even pray. For the first time in a million years, sacrificed in only decades, our fate is no longer in our own hands. Future generations turn in their graves.

At 2,000ppm, we have lost all of our reefs, the seas have risen 70m, tipping points have been triggered, the world is enduring a nightmare. War, food shortages, water shortages, disease, storms, fire, drought, flooding; lots of death. We have not cared about the planet, it is now time for the planet not to care about us. This is often called 'alarmism'; that is because it is alarming. More alarming is the political resistance to the facts. The Climate Covenant includes us all. You can choose whatever answer you like, so-called skeptic or a so-called warmist, no-one is excluded. It has to be that way. We do not all have the same values. This has to be your choice, your morality, whatever that is.

Out?

So, that is the story. This planet can hurt us severely from the burning of fossil fuels. We ignore that truth. The irony is that we do not have to take any risk; clean, cheap alternatives exist now. This is a choice between a world of health or one of pain. Politically, we have little choice or say. Our laws are corrupted. We can only change that if we believe that exponential CO2 rise is important enough to take steps to change it. If we take this first step, then this book and this process will begin our fight.

Let's also be clear on what this fight is. It is not fatalist, so do not be fatalist. Fatalism is one way of avoiding thinking this through and is itself an excuse. We cannot afford more excuses nor more lack of thought. We can succeed if we want to, it comes down to choice, not to any reduction in standard of living. If from this book we can learn how bad things are and how waiting time has run out then trust that the next can show how we can overcome the major problem, our lobbyable political system. It is the time to wait before acting that is at issue, not time itself. There is hope and hope depends on our understanding. It can be solved, but we have to be clear on how it can be solved. It cannot be solved without also addressing our corruptible political system.

It is not a technology issue, the solutions exist. It is not even that the solutions are expensive, they are not. Yes, you can complain about how they look, but they do not kill off humanity.

If you want something to do to help right now then the most important thing that you can do is to discuss this with others. Help others to realise what is happening. Yes, if you can, install clean energy, buy a clean car, install a clean heating system, install efficient light-bulbs and so on, that will all help and will make a difference. But never lose sight of the fact that that is akin to a few hairs on a very hairy head. It is not enough. Without industrial change set at a political level, it is not enough.

And yes, it will take time to address the politics. Is that time we can afford? Well, what choice do we have? If it takes a decade to solve then so be it, we had better begin today. The alternative is fatalism, that everything is lost and nothing can be done. Never think that way, it is an excuse. And that man in the mirror can make that change, as a wonderfully talented man once said.

So, the Covenant is written, the agreement is the level of CO_2 that we do not wish to breach and the year in which we want a plan to start in order not to breach that level. That to be honoured. Try to read and read again until you are certain of your answer and certain to remember it. If we forget this, then we truly forget too much. We are here. This is our responsibility, yours and mine. Learning these facts turns our innocence into reality.

19 REGISTER YOUR CLIMATE COVENANT

Well Done! You have made it through. Remember your answer; your CO2 limit and your start year. They are your contribution to the Climate Covenant and they can be registered at www.theclimatecovenant.com. It is important that you do. There are more steps to come and more to do. More care, more insight, more thought, more understanding and more helping. No excuses. Solving the CO2 problem is likely to need us all to focus upon it not for a few days but for many decades to come.

You may be wondering where the huge swathes of references to supporting papers and evidence running back over decades are, well they were left out of the paper text, on purpose. Though that does not mean that they do not exist, they do. The intent of this book is exactly to present the truth, evidence and facts as they exist within the science today. And if you do go on with this, you will find that there is also a lot of disinformation out there though now you should be able to spot which is which.

The other reason is that whilst the existing evidence changes little it is being added to daily, revealing more and more as time goes on. Even if a huge amount of references were presented today, there would be new evidence missing by tomorrow. So, I have decided instead to include them on the website at www.theclimatecovenant.com where you can read and see videos to your heart's content and the references can be kept current. It is also very much easier to click on an on-line reference than to copy the paper one in. And, if you so desire, you can become an authority on every detail of this subject, in order to explain it all to every interested party that you meet. I hope that you do.

Even if there is no end in sight, there may be a slim chance of a beginning. Words are not enough. There is hope but only if we care enough to act for others before ourselves. As you act, so will others. As you try, so will others. You will not let them down and nor will I. If you want others to help then talk to them. Keep the Covenant introduction cards with you to give to friends, family and colleagues when you see them. That is the most effective way to introduce the Climate Covenant to others. If you do not spread the word, who will? It is now up to you.

There's lots to do in very little time. It may be safe to fail at the daily crossword, but in this we need to succeed. This is neither sales pitch nor fantasy. People will die and suffer depending on what we choose to do today, and for decades to come. This is as much your responsibility as it is mine. Well done for getting this far; now make it count. Don't stop until our work is complete. Let us be blunt, save us or damn us, that choice is yours.

How You Can Help Now:
1. Go and Register Your Answer.
2. Use Your Intro Cards, Home, Out & Work.
3. Support The Next Step, The Votocracy.

Thank You,

Good Luck,

&

Until the Next Time,

The Human.